Selected Scientific Theories

By

Idegu Ojonugwa Shadrach

Dedicated to

Engineer Solomon Enemakwu Oguche

And

Engineer Joseph Prince Etim

Table of Contents

Force

Force is a phenomenon gotten from the emergence of air by the distance and measure of the element plugged into observation. The weight of the element and the present amount of air determines the acceleration between the observing points to the experimental points. What are mostly considered in force is time, weight and distance. This is the primary measures as the three of them are the major genesis and foundation of force. In a dry land, the force is secondarily considered by the slope, hill, valley and loopy level of where the test can be put to test. By see, the secondary consideration is the depth of the water, direction of the movement, and the height and length of the object or the element put to test on it. By air, the secondary consideration is the height and the length of the element and the direction plus the availability of air by the very moment. The tertiary determination of force is the controller of the element, the loads in the element and the length of the element. The safety of the movement is the level of acceleration put to consideration. In some occasions, climate changes do affect the effectiveness of the service of the elements. Therefore, the gravity of force is the total amount of distance and the weight of an object.

Time

This is the prime demanding factor for any action. It is only being which never have controlling unit towards the fast-forwarding and back-warding power. It is the essential measure of action. It ever counts forwarding and never back-warding and that make it the most rigid phenomenon on earth. Everything under the sun, being humans and natures are constantly affected with time. All creatures honors and celebrate time as it is the strong being behind the development of the world. The attractive and the damaging level of all creatures are channel through the influence of time. That is why the factor of time is inevitable in humanity. The beauty of time is gotten out of a speed put in place to arrest a particular phenomenon. If patience is given to time, bigger result can be extracted. The good measure of time is adequacy and accuracy in maintenance of a phenomenon and in generating a phenomenon. Time work beautify with climate to weaken metal and strengthen metal base on the favourable and the unfavorable stand with the particular metal. Time also work with temperature to increase and decrease active content of a chemical and where chemicals are duly applied. Time and balance diet bring effects to the growth of human beings and plants. The beauty of oxygen and hydrogen is solely determined by the possible amount and the availability of time. All positive and negative effects of life are the genesis of the artwork of time. The environmental and climate factor of time directly and indirectly affect the growth and development in humanity as the health of creatures are measured and tagged on it sake. Therefore, time create joy and sorrow. So, the dignity of the earthly activities is determined by the fate of time.

Battery

Battery gives power and external power to any scientific movable devices. Battery comprises of metal and chemical particles. This charges negative and positive movable devices to avoid future failure of discharging purposes. This happens to any movable devices that transport any substance perpendicularly, horizontally and vertically. The combination of those particles of different compound form a cell and the cell comprises of neutral which regulates the work of the negative stand to the positive stand, the positive stand that regulate neutral stand and negative stand to produce active effects and as well as the negative stand that regulate neutral stand and positive stand to produce active effects. Power is generated by the both and the reason for the negative and positive stand is to have effective saving power as negative and negative can't work effectively as it doesn't have challenge and reactive purpose and so as to the positive stand positive stand. It is designed in capacity to meet the standard of the particular devices and purpose created for.

Machine

The safety and the dangerous level and measures of humanity is kindly determines by the machines. In addition, the growth and development of humanity is also at the ground of machines. As we have different machines by the help of continues development in science and technology, so as they are designed by different purpose. Some serve special purpose while the other serves multiple-purpose. Different observations and the place of the observations helps to get plenty machines. Some machines work perfectly for their initial designed purpose while some work partially-perfect in their designated purposes. By trials, some machines are built and by adequate attempt some are built successfully. Their relevance is measured by the negative report of an element created for as machines would have gained no relevance if everything is at the perfect point or humans do not quest for more knowledge or the natures have not developed treacherous mind with one another or humans. Adequate guidelines are attached to the use of every machine to avoid rough and smooth accident. The higher humans are careful to the use of those machines in their different places. The advent of some sophisticated machines have enabled the scientists and technologists in their different areas penetrated to other planets, above and beneath. Some machines are produced to regenerate or generate other machines. Presently, the world is now getting satisfactions in almost all areas by machines. Humans have engaged in lesser work due to the machines. Manual filtration of chemicals to serve plenty tasks, rebranding of metals to serve additional tasks, beautifying the environmental disaster and putting humans in their desired shape and putting natural resources to the desired taste and form to humans are duly appreciated by the work of those machines. Therefore, the safety of the machines is the safety of humans and other creatures and the safety of humans and other creature is also the safety of the machines. Machines are the survival of human fitness. All machines are created from the natural phenomenon. Therefore, the existence of all scientific machines is the amount of natural concrete elements and human abstract elements. Time and climate are the serious determinant of the contents of the machines.

Animal

Animal is any being that can breathe, care for other beings and regenerate it kinds. Animal is bound with happiness as it is the only phenomenon that can reason that can consciously recreate other useful things in life and can recreate their kinds through sexual practice. The aquatic animals have their survival from the terrestrial animals while terrestrial animals have their survival from the aquatic animals. As they have conscious, they have choice, taste and desire and for that, some of the terrestrial animals have their survival on their fellows and others on their opposite fellows and the same thing happen in the life of the aquatic animals. They are jealous being simply for having desire. So, they do experience fight in order to satisfy their various desires. The terrestrial animals have better contents than the aquatic animals as thy do enjoy most of the relevant natural resources directly. Though, the environmental measure of the aquatic is bigger to the terrestrial but his surviving activities are put in place in terrestrial than the aquatic. The earth comprises of dry land and water land is furnished by the terrestrial animals while the aquatic animals destroy the artwork. The aquatic takes things back-world while the terrestrial take things forward. The activities of the world is generated, extended and perfected by the animal. The world is under their care. Majorly they create and destroy at instant of their choice by their power. Animal are categorized under species as they do have special characters and power in accordance to their family category. The life span of an animal is channel towards the action and reaction with other creatures. Because animals are very deadly to one another and animals are deadly in connection to other creatures. Therefore, their breath is determined by the environment they are and what they do with others as the breath is as a result of attitudinal result gotten from relationship with other animals and creatures of other species. Animals are diabolic, devious, and flexible in action. That is why the colour of the world is their architectural designs. They can live by other creatures only and other creatures can only live by them as well. Time and climate change affect them and their works. Animals are product of cells and cells is the combination of living organisms. Plants have some similarities with the animals' characters

Chemical

All chemicals are gotten from the natural resources. Chemicals are further categorized in advance level where we have solid, gaseous and liquid. Solid are gotten from rocks and mountains, gaseous are gotten from soil and water while liquid are gotten from animals and plants. Some chemicals are independent of one another while some are not. Time and climate affect the contents of chemical. It weakens some and strengthens some. Chemicals are primarily used to decorate the color of the world and everything in it. Some have long life-span and some don't. Some are poisonous only when it is mixed with other reactants, some are poisonous even when stand alone and some are not poisonous either there stand alone or mixed with other substances. Some chemical perform special work while some perform multipurpose work. Some chemicals reproduce while some do not reproduce. Some chemicals produce strength to other chemicals, some weakens. Some chemicals when mixed and some either mixed or not there can't strengthen nor weaken. Majorly, chemicals and humans are super-reactive and so as chemicals and other creatures. Only chemicals from human are merciful. Time and climate do affect chemicals. Chemicals are used to provide most of the medical solutions.

Calculation

Calculation is the primary draft of any scientific work. It is widely acknowledged that adequacy and accuracy of any scientific work is the measures of calculation. This manages effective factors and effects of time in producing any scientific inventions and discoveries. Calculation is equal to idea and time and method of any scientific revolutions. Wrong calculation leads to delay in productions and the environmental harmful effects. Right calculation leads to quick in productions and the environmental maximum safety effects. Therefore, wrong calculation equals to inappropriate methods plus and ideas all over time while right calculation is appropriate methods plus ideas all over time. In the first stage of calculation, the advantageous and disadvantageous measures of the prospective work is evaluated so as to know how to appropriate the application system of the work where and when necessary and the second page of calculation is to identify the various special and multipurpose work to avoid inappropriate service delivery that can bring harmful effects and the last most important page of calculation is the power of the prospective work to provide further learning platform and reproductions. Some calculations are inbuilt, some are out-built and others are referential.

Principle

This is a proven fact that encourages any work of relevancy. Any concrete and abstract being that do not have this certainly suffers uncertainties few moments from its existence. Principle is a surviving guardian to anything that likes outstanding results of the desired. It is highly regarded as the strong element or pillar behind any foundation. This enables to know the certain entries and boundaries of anything. If this is not gotten, useless and meaningless work can be done. Some principles are rigid while some are flexible but all depending on the position placed in the artwork. Some are emulative and some don't. Some are powerful to recreate the similar kind and others kinds. There are some elements and substances powerful enough to change principles. Principle is equal to total amount of desire plus the reasons all over the time.

Law

This is the collection of proven theories, results, observations, experiments, principles and procedures in scientific domains. This gives rise to other studies in the particular or related areas. The maintenance of laws in scientific grounds is the right and easier way to get a solution over the processing problems. In the work of science, law is rigid but can be compared to see the better, simpler and topic-oriented one. Law is propounded by the group of scientists that are either scholars or experts in the profession or in the particular topic as a theory is propounded singularly to meet the demand of the topic before a group of prominent people in the area can prove and after meeting up with the necessary demands, it can be voted as law. So, law is the final unit of learning in a particular topic. Law is analytical, comprehensive and exponential in nature.

Blood

This is the surviving community of human system. It is the control unit of stimuli. This makes endless movement to keep all cells awakens as during the movements, it supplies strength to all human joints. Life is lifeless without blood while the body decays without oxygen and the manure cannot effects without hydrogen. The heart produces blood from the absorption of solid and liquid minerals taken and the heart send the different blood to their duty's' positions. The heart pumps higher blood whenever the heart is not at maximal rest and due to the overflow of the blood from the heart, it might get finished if it is not immediately regain as that might cause consciousness and if appropriate medical care is not supplied immediately, the oxygen might finish as it do no longer have play-mate and host, and to that regards, life can be lust. If a blood meets with spermatozoa, it forms cell which will later as a result of body heat, will develop a living organism and with time, the living organism develops to become a kind of the host. The level of blood gained in the body is determined by the balance diet. The good content of human blood is the stronghold of the human body system.

System

Human system is the total amount of living organism in human body. That is the reproductive system, respiratory system, the excretory system, the digestive system and the stimulative system. These are the living organisms in human body system and there are responsible for special and multipurpose work. Oxygen and blood can't have work to do without these. Due to environmental and climate factor, the body system of humans differ as body system is highly adaptive in nature. And this is the core reason why we have different symptoms of diseases and different ways of treating them as treating a patient is best appreciated using the patient's body system choice and level. Balance diet and climate change that affect the effective contents of blood which stand firm to fight with any kind of diseases that want to gain entry into body system should be taken good care. There are some foods that are not blood-oriented and there is some oxygen that is not blood-oriented, so, disease can penetrate into the body system through these.

Matter

Matter is metals as it kindly does with things that have weight, lent and height that can occupy space. Metal is an attribute of weight, length and height as matter is the total measure of those things all over the space. Therefore, matter and metal are synonymous in scientific arena. So, as matter is an abstract and metals concrete, let us treat the both as one using collective phenomenon. We should then use length plus height all over weight. With this, we get the appropriate measure of a matter or metal. Some metals or matters have big length and height but have little weight, some have small length and height but have heavy weight, some have big length and small height and have big weight while some are the same but have little weight and some have small length and big height and have big weight while some of the same do have little weight. From there we can depict that, what matter most in measuring metals or matters is the content of the metals or matter plus the background of the metals or matters all over the space. Some metals are regenerative and some are not. They have different patterns of treating them. There are highly affected by time and climate in discharging their different duties. Some have special duty while others have multi-duties. Metals are used to do the most sophisticated weapons and machines in humanity. Only metals have direct and immediate power to destroy humans and plants of a higher numbers at a spot. Metals are decorated by chemicals to have better look. Metals are maintained by chemicals as well and the lifespan of some metals are determined by the application of some poisonous chemicals. Metals provide most of the technological solutions rising from building of the scientific works and to supporting them as well. The standard of metals is drafted from the space there will stand on. On that ground, soil types and rock types is the defining factor of metals resting space. Metals are dependent of one another. Metals prefer double functions at a goal. There are very relevant in humanity as any electrical devices are drafted and some fully gotten by metals.

Magnetism

The attraction of similar metals processed with some chemicals call for magnetism. This happens to fulfill a purpose. This calls for stronger bound of two metals to serve as a pillar and to withhold some lighter objects. It primary purpose is to attract hidden objects to the open surface.

Air

This is the composition of oxygen and hydrogen. The movements of plants and animals and other natural resources are the existence and flow of air. The direction of air is the determinant of it ventilating effects. Therefore, poor rotation or movements of natural resources and wonders bring poor air while good rotation and movements of those things bring good air. So the movement of the earth is the genesis of air. Free space also creates air. Air regulates heats and this helps comfortable environment. Adequacy of space generates air.

Heat

The movement and rotation of non-living things generates heat especially chemicals and metals. And the space occupied by matters does generate heat. The movement and congestion of living things also generates. Blockage of space does so as well. Gravity of force and distance acceleration brings heat. Pressure between two inanimate objects generates heat. Heat regulates air and makes the environment uncomfortable. Inadequacy of space creates heat.

Soil

This is where oxygen and hydrogen trades for the survival of humanity. This is a smaller part of the world where higher adventures take place. This is the only place human beings and some non-water-oriented animals live. It is hanged on water. It produces seeds of all kinds out of the dissolved manure. Plenty wonders are found on it and some are associated with it. It purposefully generates heat where there is no adequate space between humans and animals or the wonders.

Sphere

This is the determinant factor of the world. The world of harvest is kindly associated with this. It is primarily divided into lithosphere, hydrosphere and atmosphere. Atmosphere deals with the air, hydrosphere deals with the water and lithosphere deals with soil. These three things are the fundamental host of life as the three combinations produce food for human consumption. They both serve special purpose and multipurpose work. The three are dependent of one another. The earth is made up of the three. The soils when helped with water produces plants, the soil which is also known as land when helped with climates it produces temperatures and when the three are put together for help produce humans and animals. One among them can't survive without others. It is only the sphere that has similar power with time as the sphere is the channel which time is known. Time also have equal power with sphere as time do segments sphere to the recognizable form. Therefore, time and sphere have exact power and this put the both in the relevant domain. The maintenance of humanity is the artwork of the both. The world is ever submissive to them as there are the wonders of our Creator. The growth and development of the world is treated accordingly to the wish of the both. Water gives life, soil creates life and the climates maintain life. The bother create problems of life and offer solutions whenever their creatures go wrong with one another. Chemicals and metals work in accordance of climate desires. Chemicals and metals always give problem to the climates and thereby allowing climates descend notoriously on humans by air pollution and whenever the climates have finished with the chemicals and metals porously, the soil absorbs them while the water gives them second chance and change. All there produces especially humans regulate their handiwork and make changes to their original definitions. Other creatures of their power can control them minimally but only humans do it in a maximal level. So, time and occasions, define who there do betray and trust themselves. But all that is at unequal level as abstract and concrete beings never share equal advantages and disadvantages on one another. The both serve as manure to one another.

Rotation

This is the movement of earth from one unit to another. It is caused by time and climates. The channel of the movement is scientifically known as the latitudinal and longitudinal movement. The rotation is not physically noticeable. It is a gradual move and as it revolves so as climates changes. Whenever new or strange climate change occurs, it is an identification of the rotation. It has little effects on living and non-living things. The sunrise and sunset and direction of air and the direction of rainfall identify the rotational effects.

Method

This is the scientific direction in making a work done as it is the processing unit of any finished work of science. This has various types. Appropriate selection of its type to deliver a desirable service is the relevance of its existence. This is the prevailing record of any scientific breakthrough. This is a channel of reaction between raw and un-raw information. The existence of this creates plenty artwork of knowledge. This also gives different meanings to learning platforms. Method is equal to the total amount of observation plus the total amount of processes all over the time used.

Climate

This is the complete sphere of atmosphere. It is the higher ground for polluting the world. The climates absorbs and dissolves every temperatures produced by the activities of the world. Climates have the largest determination of human survival on the surface of earth. The changes of the climates create drastic changes of the world.

Distance

This is a space between two axes. The cover is entitled with speed plus velocity all over time. To obtain any goal, there is a necessity for distance as if everything is around; there is no need for struggling. Struggling to create things bring the relevance of distance. The release of any scientific work is placed between the gab from the starting point and the ending point. Therefore, the middle point is the determinating factor to either strengthen or weaken the work or test. The factor of longitudinal distance is the slope of the land while the latitudinal distance is the amount of the available air. The length of the distance determines the time measure. If a distance of anything is covered, it secrets are known. Distance is equal to the total amount of the starting point to the ending point plus the total amount of load delivered all over the time used.

Absolvation

This is a process of combining any of the three against one another or against mixed combinations of the same and other chemicals (liquid, gaseous and solid). There are put together to generate new kinds or the related kinds for a special purpose or multipurpose work. Resultant of some absolvation is high poisonous, some are partially poisonous while others are not poisonous at all. Major reaction takes place in water when two chemicals of single compound or two or more chemicals of different compounds are suffocated in water. In some occasions, the water reacts with a single chemical put into it. Generally, it is when the water is applied to any chemical before it can have power to react. Colours and contents are changes during the absolvation. The time is the major determinant factor of colours and contents changes in absolvation. Therefore, absolvation is equal to the total amount of water and total amount of chemicals all over the time. Temperature is the secondary determinant of observation while humans are the tertiary determinant of absolvation.

Observation

This is the total collection of test, attempts, gathered information and points gotten about a particular topic in scientific area. Observations prompt new and further study. Good observation brings good process to finally have good result while wrong observation brings wrong result after being processed. Therefore, observation is the bedrock of scientific works. Open observation and secret observation are needed to consolidate a fact as through the mixed observations, there are some hidden things that will be open or some open things which are due for reshape to give better results concerning the observations. Observation is equal to the total amount of information plus the total amount of methods applied all over the time.

Result

This is the final level of any scientific work and learning platform. It is the platform that establishes further studies in the particular topic or related topic. Result is known as the solution of a problem. This modifies and qualifies any topic of choice. And this is the only reason to undertake or process any problem. It is reality-oriented and idealism-oriented. This only gives courage to any learning or problem. We have special result and multiple purpose results. And result is equal to the total amount of problems plus the total amount of processes plus the total amount of methods all over the time. All elements of result is essentials, the components and particles as well. Result is also known as solution in scientific floor.

Problem

This is the single phenomenon that calls for the attention of learning. Without learning there will be no strive for perfection. The world of knowledge would have remain dark without existence of problem and only problem creates room for competent arguments, observations, methods and processes to be responsible enough to create result as the excess of problem is solution finding. We have special and multiple-purpose problems. Problem is the total amount of issues plus the total amount of processing method all over the time used.

Water

This is the only resources that perform all surviving work. Water and time changes all things on earth. Their power is as good as life and death. It is noted that water gives life. The survival of all living organisms is rooted in water absolvation. What changes climates, chemicals and other creatures of the earth.

Sound

This is produced as a result of an accident. If two or more metals meet unconsciously, sound is produced. The same thing happen when chemicals are dangerously mixed. We have natural sound of the climate. As humans interact or engage in any action, sound is produced. The same thing happens between two other animals and plants. Sound is filter in some occasions to attract good hearing or voice. The major reason for sound is to communicate against the previous or present or future things. Therefore, sound is the total amount of pressure plus the information all over time.

Filtration

This is a process of changing the content of a particular chemical to another. It equally changes chemical from one form to another. This creates essential demarcation that gives beauty to any scientific work. This in some occasions does with withdrawing useless substances or organisms to bring original beauty of a chemical. It is therefore an agent of purity. It is a scientific device and also a scientific process as enhanced by various place of need.

Step

This is the serial occurrence of process in ascending or descending form to produce reasonable and expected result. It is a medieval between problem and solution. This is a technical point of any scientific work. Because, once a step is wrongly indented, the whole work is under destruction. As scientific work deals with all areas that affect human life directly, it is very important to handle a step of everything carefully to avoid dangerous happening before, during or after the work delivery.

Experimentation

This is the test of the overview of the observation. This paves better ways and rooms for analysis. This is the host of decision on how the practical work should be done for the purpose of the topic's safety. The channel of plenty filtration and steps will be put to test in order to see the advantageous and disadvantageous level of the study before concluding where and when necessary it is to be used. Possible amendment is made to any practical work during experimentation. Experimentation is therefore equal to the total amount of analysis plus the steps all over the time used. Experimentation prompt plenty results and identifies plenty problems. So, high level of intelligent is used during experimentation.

Erosion

This is the accidental joint of two natural phenomena on wonders. The inappropriate and inadequate space created by soil and water bring erosion. The types of particles that are manured usually determine the direction of erosion. It is natural as that entire bound together to do this are wonders of God. It can only be controlled by humans but can never be protected. Erosion is of two types. The second one is of the man-made phenomenon. As erosion is the division of a flat space into two by the pressure of the environmental accident, so as building and other works that is divided into two because of external force of inadequate and inappropriate joint, is simply that. The protection can take a form of primary, secondary and tertiary.

Accident

This is the unconscious combination of two substances either by similar compound or different compounds. Wrong connections of metals and chemicals are equally known as accident as the result is dangerous or harmful. We have conscious accident which is the combination of two similar or two un-similar substances from either one compound or different compound on a purpose of learning to see result it might bring. We equally have external and internal accident and as the words connote, there are some accident that take place secretly form the wrong and conscious combinations that do not immediately react but has power to gradually decreasing the content of the result of the successful combination while the external is the one that has the reaction meant immediately after successful combination. Accident is therefore the total amount of wrong calculations plus the total amount of the elements all over the time used.

Pollution

This is unfavorable atmosphere caused by natural accident or man-made accident. This is the genesis of diseases and spread of diseases especially the epidemic diseases. Pollution can be minimally controlled and prevented.

Vacuum

This is a channel of communication between two or more scientific devices to the destination where necessary to perform the specific work. It is a gap between tow active and inactive elements or devices and organisms. The vacuum determines the strength of work to be performed. All secret activities of a device take place at the vacuum arena. The vacuum is the mode of calculation of any scientific work.

Sun

This is the determinant of world activities. It affects all creatures on the dry-land and on the water-land positively and negatively depending on the right and wrong time of the actions. It is the general light of the world. It exchanges position and time with moon at due moment to honour the universal law of day and night. It partially gives time to rainfall and it fully controls the affairs of photosynthesis and evaporation. It is located at the center of the world and that is why the direction of sun looks vertically, horizontally and perpendicularly to the position of all humans.

Practical

This is the raw assessment upon an observation. This is where and when major work is done. This gives credit to the subject and gives relevance to the theory. The concluding part of practical gives birth to suggestions and contributions before it can be finally recommended. This is the technical part of learning and this is simply the reason of learning. So, practical is the most essential part of learning.

Device

These are the transferring medium from one point to another in the scientific fields. There serve as median elements that connect two bigger elements for adequate and accurate functions.